Teach Me!
Direct Current
(Beginner)
M D Jubenville

DEDICATION

To my Lovely Wife, who has always had my back,
even when I had no idea where I was leading her.

CONTENTS

ACKNOWLEDGMENTS

I would like to acknowledge my **parents**,
who put me through school for
Electronics Engineering Technology;

I would like to acknowledge my **family**,
who were there when I became
a System and Software Developer
creating my own consulting business;

I would like to acknowledge my **wife**,
who sacrificed to get me through school for
Electrical Engineering Technology.

CHAPTER 1

Who should read this book?

Teach Me! Direct Current (Beginner) - this book - was written for anyone who wishes to start learning about Direct Current (DC) electricity. The basics are covered without a lot of focus on the physics and theory.

Electricity always involves math, but only the basic equations will be discussed as well as Passive Components and a teaser for what is to come in the other books in this series.

Teach Me! Direct Current (Intermediate) dives into the physics behind electrons and electromagnetism and looks at the time constants for passive components. Semiconductors are discussed in detail, including how to power them, use them and understand what they are doing. Charging and charge-control circuits are discussed.

Teach Me! Direct Current (Advanced) discussed electromagnetic devices, including relays, speakers and microphones, and touches on communication. Learn about all types of amplifier circuits and how to use Op-Amps. The last part of the Advanced book touches on Alternating Current, mostly with respect to small-signal analysis.

CHAPTER 2

What IS electricity?

Simply speaking, electricity is the flow of electrons, usually referred to as electrical current. You may hear or read the term **Conventional Flow**. This describes the flow of electrical current from positive to negative.

Conventional Flow

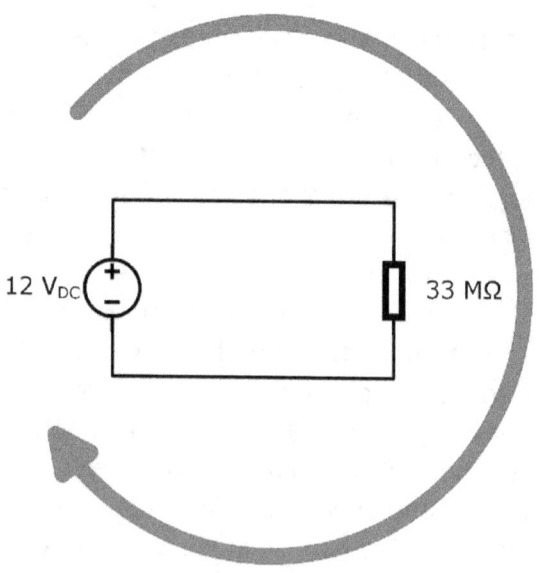

Electron Flow, however, describes the actual movement of electrons from negative to positive. For simplicity, electron flow will be used when discussing electrical current in this book.

Electron Flow

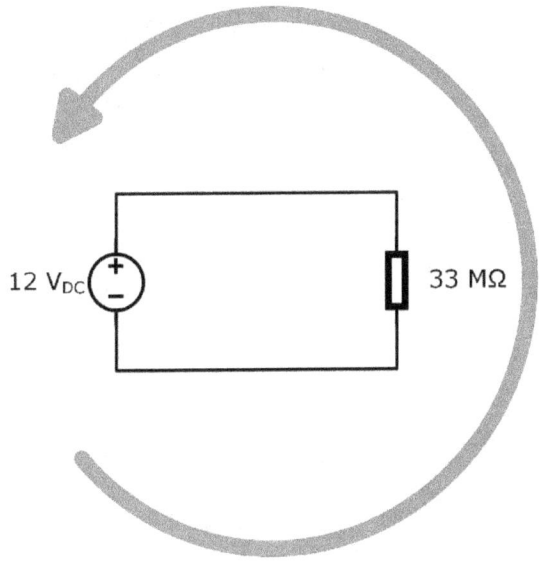

The electrons that make up the electric current are the free electrons found in conductive materials, or conductors. Copper is an excellent conductor, because it has one free electron in its outer shell or valence shell. Think of it as the electron's "orbit". This free electron can be coaxed into moving from its own atom to the next one without much provocation – or in electrical terms, potential difference. Most good conductors have one or very few electrons in their outer shell.

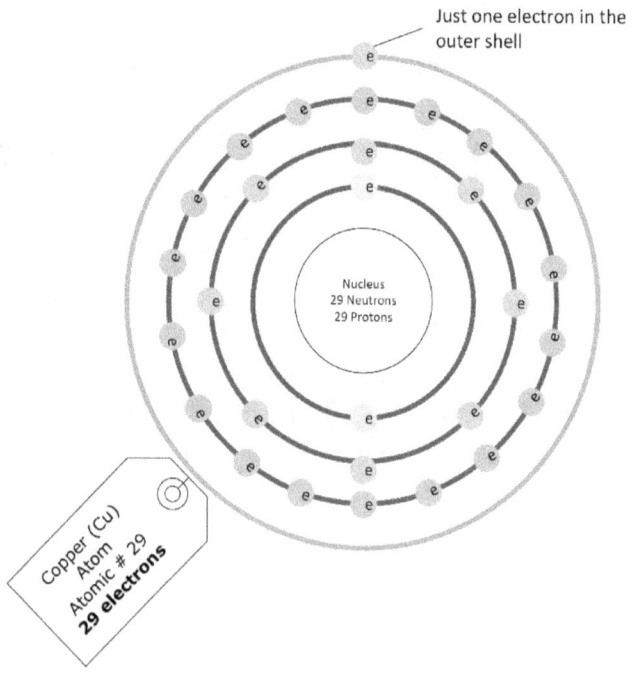

Just one electron in the outer shell

Nucleus
29 Neutrons
29 Protons

Copper (Cu)
Atom
Atomic # 29
29 electrons

Insulating materials, or insulators, have either very few or no free electrons; they usually will not allow electricity to move through them. Their outer orbit or valence shell is close to being full, or is full. Copper wire, a good conductor and reasonably strong, is usually wrapped in an insulator of either a plastic-like (thermoplastic) or a rubber-like (thermoset) compound.

The image shows two larger XLPE (Cross-Linked Polyethylene) wires and three standard PVC insulation wires. The XLPE is being used for the high-current, high-temperature characteristics for the insulation.

Microphone cables use thermoplast-sheathed cables, for example, which provide more flexible insulation.

CHAPTER 3

What is Direct Current?

Direct Current (or **DC**) is the name given to the flow of Electricity in only one direction. That is the basic definition of direct current. Electrons flow through a circuit. There may be multiple sources of current (batteries, power supplies, devices) and multiple paths for the flow to follow, but as long as the current flows in one direction, it is considered Direct Current.

You may have read that DC is a constant voltage but that simply is not true. If the direction in which the current is flowing does not change, it is DC. If the current reverses – where **positive** becomes **negative**, and **negative** becomes **positive** – that would be classified as Alternating Current (or AC) but that is covered in the **Teach Me! Alternating Current** series of books.

Voltage

Water & Electricity – Analogy for Voltage

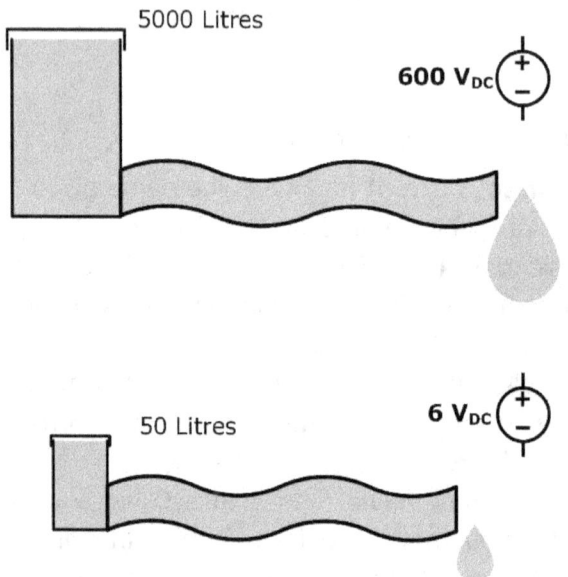

Voltage potential is very similar to water pressure.

A water tank is similar to a battery. High water pressure is like high voltage "pressure" or potential difference (which can just be called potential). Low water pressure is like low voltage potential.

A small tank with high pressure can shoot water through a hose quickly, but it does not last very long. A small battery with a high voltage can shoot electricity through a wire (or conductor), but the battery loses its charge in the same way.

A large tank of water with a low pressure might

trickle water through a hose, and a large battery with a low voltage will trickle current but will last longer.

The higher the voltage, the higher the electrical potential. Generally speaking, the larger the battery, the longer it can hold its charge. That potential is what pushes electrons from one end of the circuit to the other, just like high water pressure pushes water from one end of a hose to another.

Electrical potential or **voltage** is represented by the letter **E**, and is measured in **volts**, whose symbol is **V**. Often, especially if there are different types of voltages in a circuit, DC Voltage is written as V_{DC}.

Batteries are a common (and well-known) voltage source for DC. Almost everyone is used to using batteries of nominal voltages: 1.5 V cells (A, AA, AAA, AAAA, C and D), 6 V "Lantern" batteries, 9 V cells (PP3) "Radio" batteries and 12 V "Car" batteries. There are many more types and voltages for specialty batteries, such as watches, hearing aids, car fobs, laptops, tablets, and phones which range between 3 V and 30 V.

Not all systems that are DC are battery driven. Chargers for your phone, Kindle, rechargeable flashlight are generally plugged into an electrical outlet (this is AC in most countries) and that is converted to DC. Outdoor residential lighting is often 12 or 24 volt DC, powered again by an AC source. Many appliances and home electronics are plugged into an AC source, but most of what is inside is powered by DC control circuits. Most electronic control systems in factories and vehicles operate on 24 V_{DC}.

There are DC sources that have nothing to do with plugging into an AC outlet. Solar cells, connected as solar panels, are becoming quite commonplace in both grid-tied and off-grid systems. Grid-Tied systems change DC to AC and feed the electrical service grid just as power stations do. Off-grid systems use local storage in batteries and power either DC or AC systems in a cabin, recreational vehicle, boat, shed and even a home or apartment complex. There are also DC generators that run on fossil fuels. Another type of DC generator is the Hydrogen Fuel Cell, which converts hydrogen and oxygen gas into DC and water.

DC is not just small. Most solar panels produce between 12 and 48 V_{DC}. Some solar arrays (many panels together) can run anywhere from 120 to 600 V_{DC}. Diesel-electric generators in locomotives product from 1200 to 3000 V_{DC} to run the traction motors underneath them. There are High-Voltage Direct Current (HVDC) installations in many countries including Canada and the United States that operate between 400 kV_{DC} and 600kV_{DC} - that means 400,000 to 600,000 Volts!

Current

Water & Electricity – Analogy for Current

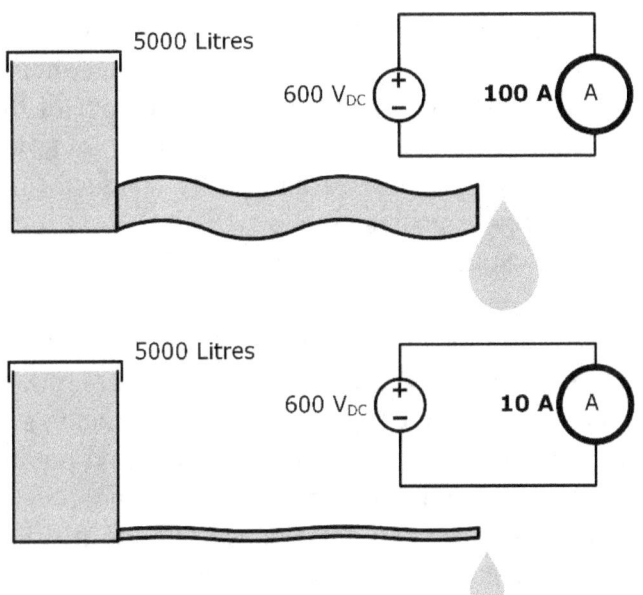

If voltage is like the pressure in water "circuit", current would be like the amount of water flowing through a hose at any given time. That flow of water is just like the flow of electrons. A small garden hose with low water pressure would have a low flow. Similarly, a small circuit with a low voltage would produce a low current. A high pressure going through a large fire hose would have a high flow. A high voltage would also produce a higher current.

The analogy to water with respect to electricity might help to explain how electrons move through conductors. The water in the tank has to travel through the entire hose before it comes out, but the water already at the end of the hose comes out before the water at the beginning of the hose which is connected to the tank. Electrons in the battery's negative terminal are flowing into the end of the connector attached to the terminal, but the electrons at the positive end of the conductor are the first ones to come out.

This explains how electricity moves so quickly through very long conductors, such as the transmission lines on towers connecting cities to power plants. The electrons at the power generators are put in one end of the conductor, and for most electrons going in, electrons come out the other end. It is not a one-for-one exchange, and a single electron in one end does not force a single electron out the other. However, it does generalise how the interaction of potential across a conductor works.

Electrical **current** is represented by the letter **I**, and is measured in amperes or **amps**, whose symbol is **A**.

Resistance

Water & Electricity – Analogy for Resistance

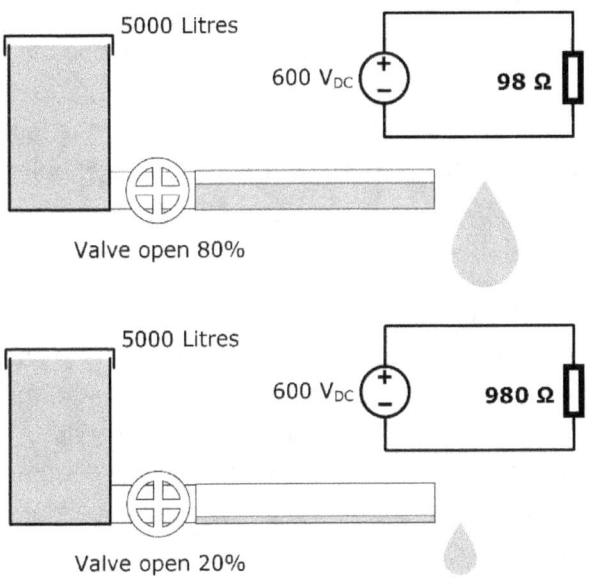

If voltage potential is like water pressure and current flow is like water flow, Resistance is a measure of the slowing or restriction of flow and electrical resistance is the measure of the slowing of current from the negative to the positive. A very high resistance will almost completely stop the flow and a very low resistance will do almost nothing to stop the flow. With water, a valve can start, stop, and control the amount of flow being sent down the hose.

Resistance in electricity can start, stop, and control the amount of electrical current flowing through a circuit.

Resistance is not just for controlling current. Every component in a circuit has resistance. Lights, heaters, motors, speakers all have a resistance to current flow. Even the conductors in a circuit will have some resistance.

Electrical **resistance** is represented by the letter **R**, and is measured in **ohms**, whose symbol for resistance is Ω (the Greek letter omega)

Ohm's Law

There is a formula that relates electrical voltage current and resistance, created by German physicist and mathematician Georg Simon Ohm.

The voltage measured in volts is always equal to the current measured in amps times the resistance measured in ohms. This means that voltage, E, is always equal to the current, I, times the resistance, R. This is written as:

$$E = I \times R$$

where E is in Volts (V), I is in Amps (A) and R is in Ohms (Ω).

$$\_\_ V = 2 A \times 6 \Omega$$
$$12 V = 2 A \times 6 \Omega$$
$$E = 12 V$$

Of course, as any mathematics formula, it can be rewritten to find current or resistance as well.

R = E / I and **I = E / R**.

R = E / I

__ Ω = 24 V / 3 A

8 Ω = 24 V / 3 A

R = 8 Ω

- and -

I = E / R

__ A = 6 V / 6 Ω

1 A = 6 V / 6 Ω

I = 1 A

Power

One could consider the measure of the volume of water the same as the measure of Power. Power is measured in Watts is generated by an amount of current going through a resistance, the voltage applied to a resistance, or the current with respect to the voltage going through a circuit.

Electrical **power** is represented by the letter **P**, and is measured in **watts**, whose symbol is **W**.

Energy

Energy is the amount of power over a measure

of time.   Electrical energy which is measured in joules is based on the amount of power over a certain amount of time.   1 joule is 1 Watt for 1 second, or 1 Ws (watt-second).

Electrical **energy** is represented by the letter **e**, and is measured in **joules**, whose symbol is **J**.

Joule's Law

There is a formula that relates electrical voltage, current, resistance, power, and energy, through what is now knows as Joule's first law, recognised by English physicist, mathematician and brewer, James Prescott Joule.

The power measured in watts is always equal to the current measured in amps times the voltage measured in volts. This means that wattage, P, is always equal to the current, I, times the voltage, E. This is written as:

$$P = I \times E$$

where power, P, is in Watts (W), E is in Volts (V), I is in Amps (A).

Because voltage E can be written using Ohm's Law as:

$$E = I \times R$$

you can replace E with I X R.

$$P = I \times E$$
$$P = I \times I \times R$$
$$P = I^2 \times R$$

Current can be written using Ohm's Law as:

$$I = E / R$$

Which means you can replace I with E / R.

$$P = I \times E$$
$$P = (E / R) \times E$$
$$P = E^2 / R$$

Energy is calculated as power over a period of time. One watt for one second is equal to one Joule. Regardless of the way in which electrical power is calculated, the number of watts time the number of seconds is equal to the number of joules. The formula suggests – and is correct – that the measure of energy can also be expressed as watt-seconds. If you have 36 watts for 100 seconds, that would be 3600 watt-seconds – or 1 watt-hour.

For example, calculating for ten one-hundred watt bulbs running for 8 hours overnight would be, using the standard form:

10 bulbs X 100 W X 8 h X 60 min X 60 s
1000 X 28800 or 28, 800, 000 Ws (watt-seconds)

28, 800, 000 can also be written as 28, 800 kWs (kilowatt-seconds) or even 28.8 MWs (Megawatt-seconds).

Megawatt-seconds is not really a useful term for energy in practice, only in science. It is more common to use kilowatt-hour when talking about energy, because that is how everyone pays for energy. That means that 10 bulbs X 100 watts = 1000 W or 1 kW. That one kilowatt (kW) for 8 hours would be 8h X 1kW, or 8 kWh (kilowatt-hours). If you divide by seconds (60 per minute) and minutes (60 per hour) you can divide 28, 800 kWs by 60 X 60, or 3600, and you would also calculate to be 8 kWh of energy.

Using a price of 12 cents per kWh, you would know that it costs $0.12/kWh X 8 kWh, or 0.12 X 8 cents, which is just less than one dollar. Four weeks of those 10 lights being on eight hours each day would roughly cost about $28.

Metric Prefixes

In reading, writing, talking or experimenting with electricity, you will use the metric prefixes often. They have all been adopted by the International System of Units or **SI**, abbreviated from the French Système International (d'unités) as measurement prefixes.

Most often, DC circuits are measured in Volts, V; milliamps, mA; kilohms, kΩ. Of course, some circuits have MV voltages, kA currents and MΩ resistances; in others, mV, µA and mΩ can be used. With capacitors, which you will read about later, common sizes are micro, nano and pico Farad – µF, nF and pF.

These are <u>not different units</u>, they are just a way

to write 47 pF rather than 0.000000000047 F and who wants to say, "zero point zero, zero, zero, zero, zero, zero, zero, zero, zero, zero, four, seven farads", instead of, "47 picofarads"?

The ones in bold are used often in electrical measurements, but all are included in the following table, as well as Appendix A

Prefix	Symbol	Multiplier	Exponent
yotta	Y	1,000,000,000,000,000,000,000,000	10^{24}
zetta	Z	1,000,000,000,000,000,000,000	10^{21}
exa	E	1,000,000,000,000,000,000	10^{18}
peta	P	1,000,000,000,000,000	10^{15}
tera	T	1,000,000,000,000	10^{12}
giga	**G**	1,000,000,000	10^{9}
mega	**M**	1,000,000	10^{6}
kilo	**k**	1,000	10^{3}
hecto	h	100	10^{2}
deca	da	10	10^{1}
1	unit	1	10^{0}
deci	d	0.1	10^{-1}
centi	c	0.01	10^{-2}
milli	**m**	0.001	10^{-3}
micro	**μ**	0.000001	10^{-6}
nano	**n**	0.000000001	10^{-9}
pico	**p**	0.000000000001	10^{-12}
femto	F	0.000000000000001	10^{-15}
atto	A	0.000000000000000001	10^{-18}
zepto	Z	0.000000000000000000001	10^{-21}
yocto	Y	0.000000000000000000000001	10^{-24}

What is it good for?

Enough with the names and theories and math. What can be done with electricity and how does it do it?

Simple DC Circuit

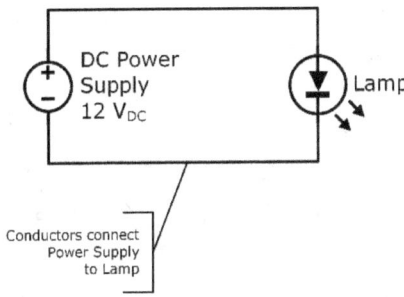

One of the simplest DC circuits (AC as well) is a voltage source connected to a lamp. Current flows as electrons from the negative terminal to the positive terminal through conductors connecting the two separate contacts to each of the terminals. That lamp is a device or component, and there are many kinds of devices.

CHAPTER 4

Devices

The very simple circuit previously mentioned is only a voltage source and a single device. That device was called a lamp, but that could be representative of any device that gives off light. Incandescent bulbs, pilot lights, Light Emitting Diodes (or LED's, which is pronounced as EL-EE-DEE's, not LEAD's, because it is an initialism, not an acronym) are all types of lamps. There are many other categories of devices, and many different types of devices within each category.

Each device has a symbol that is used when creating schematics or diagrams. There is an accepted standard from the "global" group **IEC** (International Electrotechnical Commission) which has mostly replaced the standard from the United States group **ANSI** (American National Standards Institute).

ANSI vs IEC

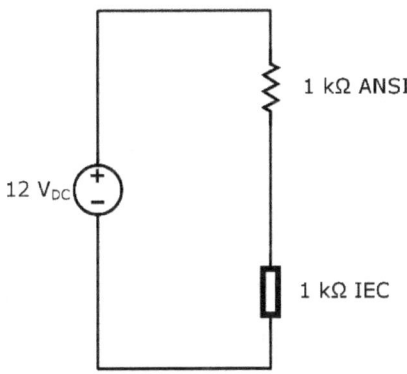

For this book, diagrams and schematics will use IEC, but there is a sample of IEC and ANSI symbols in **Appendix D**.

If we add a switch to the simple circuit, the flow of electricity can be stopped. A battery, lamp and switch are the devices in the circuit for a common flashlight.

Simple Flashlight Circuit

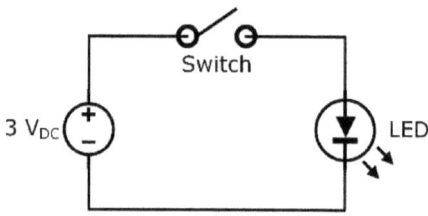

When a switch is open, current does not flow; when closed, current flows. If you want to put that into the equation, it can be considered a resistance. Open, it offers (in theory) infinite resistance. 12 V / ∞ Ω (that is the symbol for **infinity** ohms) = 0 A. Closed, it offers (again, in theory) zero resistance. 12 V / 0 Ω = WAIT! You cannot divide by zero!

That is ok, the resistance of any conductor is never truly 0 Ω, there is always some resistance. There is even resistance in batteries and in power supplies (as well as current limiters). If you have conductors made of 14 AWG (American Wire Gauge, found in Appendix C) copper wire, the resistance of the conductor would be about 2.5

ohms per 1000 feet, if you have a 6 V battery and 14 AWG copper wire, the current in your wire should be roughly 1200 A – but the rating for 14 AWG copper is a little over 20 amps. It is very likely your circuit would be a quick flash, maybe a popping noise, and it would be destroyed.

The switch is not the only device in the circuit, though. Generally, you want your conductors to be very, very low resistance. If the lamp in your circuit is a 6V, 12W bulb, you know that using:

$$P = I \times E$$
written as
$$P / E = I$$
$$12\ W / 6\ V = 2\ A$$

Now, using:
$$E = I \times R$$
written as
$$E / I = R$$

$$6\ V / 2\ A = 3\ \Omega$$

You can check that with
$$P = I^2 \times R$$
$$P = 2^2 \times 3$$
$$= 12\ W$$

Passive Components

Generally speaking, passive components are those that consume or store electric charges. The rule is debatable, however, when speaking of passive components, most people are referring to resistors, capacitors, and inductors. Electromechanical devices (switches, connectors, and so on) are not considered passive components. Some consider lamps to be passive (a light-emitting resistor); others consider lamps to be active (electric current to light transducer).

It might surprise you to know that transformers are considered passive components. They are often used to raise and lower voltage, they do not generate power, which means that they only consume or store. Transformers are a usually a set of inductors but are more likely to be used in AC circuits rather than DC.

Active components - not covered in this book - include voltage and current sources, generators, transistors, diodes and other semiconductors.

Resistors

Resistors are components that do exactly what their name suggests: They resist the flow of electricity. Measured in Ohms (Ω), resistors are often available in standard sizes. The new, tiny, surface-mount resistors are tiny components with their sizes written on them; the older-style radial resistors still use the same colour-coding system that has been in use since the 1920's. APPENDIX C has the colour-coding key.

Simple R Circuit

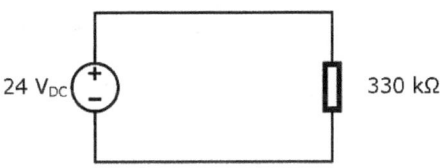

If a circuit has a 24 VDC power supply and needs to be limited to between 70 and 80 µA, it will need a resistor calculated by 24 V / 75 µA = 320 kΩ. A 330 kΩ resistor is a standard size, so that would be the size that would be used. The current would be 24 V / 330 kΩ = 72.7 µA.

If the resistor needs to be more precise, a more expensive process is used to ensure that a 330 kΩ resistor is very close to 330,000 Ω. That precision is referred to as tolerance and is expressed as a percent, meaning that if a 330 kΩ resistor has 1% tolerance, it would be as much as 1% of 330,000 (or 330 Ω) more or less than 330 kΩ. Less precise resistors of say 20% tolerance are much less costly to manufacture. It does mean, however, that the 20% resistor is 20 times less precise than the 1% resistor.   The same 330 kΩ resistor at 20% tolerance would be between 264 kΩ and 396 kΩ. Sometimes, the precision is not important, so it is ok to use a 20% tolerance resistor. High-precision resistors can have a tolerance of 0.005% - that means the 330 kΩ would be between 328,350 kΩ and 331,650 kΩ.

Capacitors

Capacitors – also known as condensers – are components that take electrical energy and store it in an electric field. DC current cannot pass through the capacitor; however, it appears to pass through as it charges the capacitor and the current drops as the capacitor reaches the same voltage as the DC power supply. If the circuit is changed so that the capacitor can discharge, the electric field drops as the electrons stored in the capacitor can flow out. This is similar to charging up a small battery, then using the stored charge until it is gone.

The amount of charge a capacitor can hold is capacitance, measured in farads (named for English physicist Michael Faraday). The farad (F) in SI units is: one second to the fourth power ampere squared per kilogram per meter squared. This is written as $(s^4 \times A^2) / (kg^2 \times m^2)$. Do not remember that, it is only written here to show the base units. One farad is the charge of one volt producing one amp per second (also known as a coulomb after French physicist Charles-Augustin de Coulomb, which is approximately 6.24 X1018 electrons). One farad is quite large – most capacitors are in the micro, nano and pico range. There are some capacitors, called supercapacitors, which are sometimes measure in kilofarads, and are often used in backup power supply applications or very large energy storage systems.

Capacitor Charge & Discharge

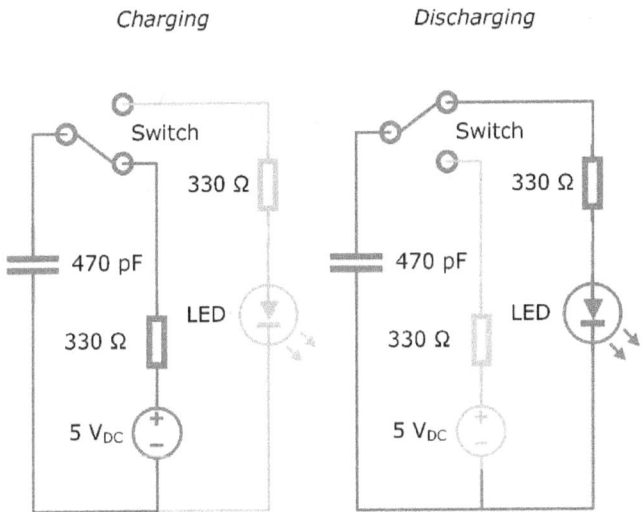

RC circuits, or resistor-capacitor circuits, allow a charge to fill a capacitor over a period of time through the resistor. That time is based on a constant, or Tau (T), is resistance in ohms times capacitance in farads. A 100 kΩ resistor with a 47 µF capacitor has a T = 100 000 X 0.000047 = 4.7 seconds (s). The resistor slows current flow, which is then collected by the capacitor as it builds up a charge. When first energized, the current will flow according to Ohm's Law. The capacitor will reduce the current flow as it charges and the voltage across the resistor will approach 0V.

Simple RC Circuit

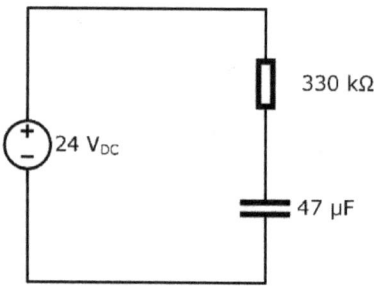

A capacitor resists a change in voltage. A 680 µF charged to the same voltage as a 68 µF will have ten times the electrical charge (capacitance) on it.

Inductors

Inductors – also known as coils, chokes, and reactors– are components that take on electrical energy and store it in a magnetic field. DC current passes easily through an inductor; however, it appears to slow as it charges the inductor and the voltage drops as the inductor reaches the same current as the circuit allows. If the circuit is changed so that the inductor can discharge, the magnetic field drops, and the electrons are induced in the inductor to flow out. This is similar to charging up a small battery, then using the stored charge until it is gone.

Inductor Charge & Discharge

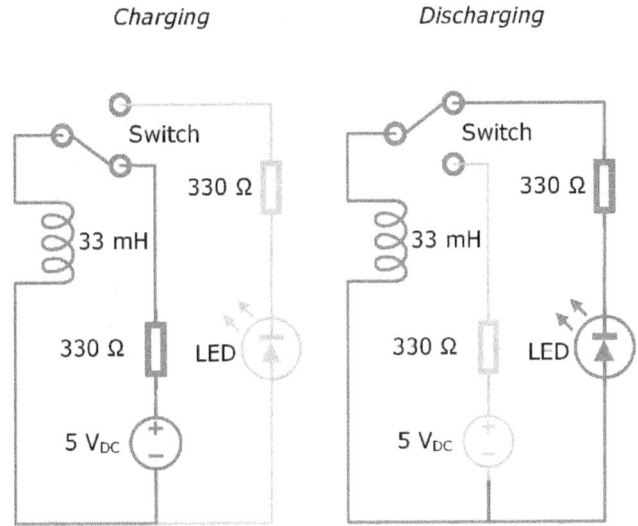

The amount of charge an inductor can hold is inductance, measured in henrys (named for US scientist Joseph Henry). The henry (H) in SI units is: kilogram meter squared per one second squared ampere squared. This is written as $(kg \times m^2) / (s^2 \times A^2)$. Do not remember that, it is only written here to show the base units. One henry is the charge of one amp per second to create one volt. One henry, like one farad, is quite large – most inductors are in the milli, micro, and nano range. There are some inductors, used for electromagnets in supercolliders, which are in the kilohenry range.

Simple RL Circuit

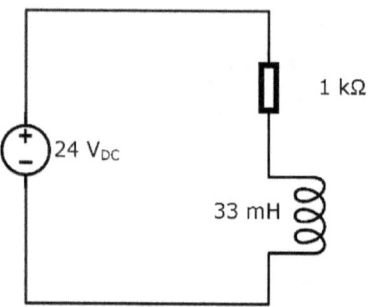

RL circuits, or resistor-inductor circuits, allow a charge to build on an inductor over a period of time through the resistor. That time is based on a constant, or Tau (T), is inductance in henrys divided by resistance in ohms. A 1 kΩ resistor with a 33 mH inductor has a T = 0.033 / 1000 = 33 microseconds (µs). The resistor slows current flow, which then induces a magnetic flux by the inductor as it builds up a charge. When first energized, the voltage will be according to Ohm's Law. The inductor will increase the current flow as it charges and the voltage across the resistor will approach the voltage from the power supply.

An inductor resists a change in current. A 330 mH charged to the same voltage as a 33 mH will have ten times the magnetic charge (inductance) on it.

Kirchhoff's Laws

The examples so far have been quite simple, so calculating voltage, resistance or current has also been simple: E = I X R. Consider the circuits below.

Two Series vs Two Parallel Resistors

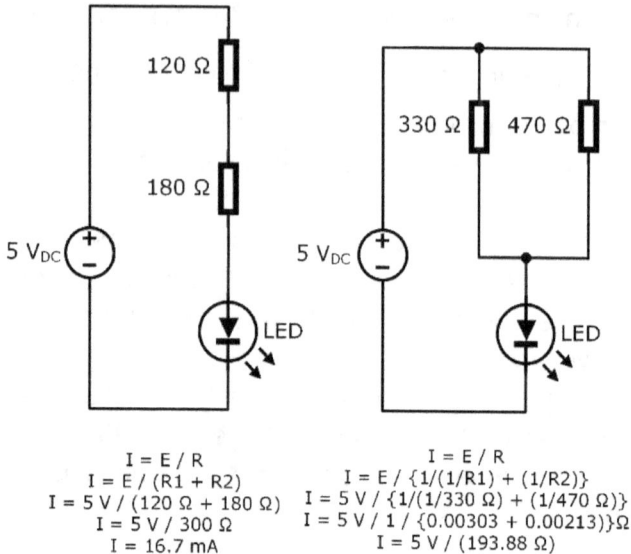

$$I = E / R$$
$$I = E / (R1 + R2)$$
$$I = 5\ V / (120\ \Omega + 180\ \Omega)$$
$$I = 5\ V / 300\ \Omega$$
$$I = 16.7\ mA$$

$$I = E / R$$
$$I = E / \{1/(1/R1) + (1/R2)\}$$
$$I = 5\ V / \{1/(1/330\ \Omega) + (1/470\ \Omega)\}$$
$$I = 5\ V / 1 / \{0.00303 + 0.00213\}\Omega$$
$$I = 5\ V / (193.88\ \Omega)$$
$$I = 25.8\ mA$$

The first shows a voltage source, two resistors and an LED; the second shows a voltage source, two resistors and an LED. Which circuit will have a brighter LED? German physicist Gustav Kirchhoff answered that question. This book is not going to get into the physics (that is dealt with in the Intermediate and Advanced level Teach Me! Direct Current books) but there are a few formulas which help calculate voltage, current and resistance of series and parallel circuits.

Kirchhoff's Current Law - all currents entering a node must be equal to all currents exiting a node. This is referred to as Conservation of Charge.

Kirchhoff's Voltage Law - all voltages in a closed circuit must add up to zero volts. One 6 V_{DC} battery connected to one resistor in a circuit will drop 6 V_{DC}. This is referred to as Conservation of Energy

Resistors in series can be treated like one large resistor: the current must pass through the first resistor and the second resistor to complete the circuit. The two resistances are simply added together.

$$RTOTAL = R1 + R2 + ...Rn$$

Resistors in parallel are calculated a bit differently. Thinking back to the water examples, resistance is the opposition to flow. One small hose would let less water pass through it than a larger hose. However, if you have two small hoses working in parallel, more water will pass through in total than with just one of them. This is how electricity behaves as well. Each resistor allows current to pass through it, allowing for more current than either resistor on its own.

$$1/RTOTAL = 1/R1 + 1/R2 + ...1/Rn$$

Inductors in series can be treated exactly the same as resistors: the current must pass through each inductor to complete the circuit. The two inductances are simply added together.

$$LTOTAL = L1 + L2 + ...Ln$$

Inductors in parallel are calculated the same as resistors, too. Each inductor allows current to pass through it, allowing for more current than either inductor on its own.

$$1/LTOTAL = 1/L1 + 1/L2 + ...1/Ln$$

Capacitors in series can be treated exactly the

same as resistors in parallel.

$$1/CTOTAL = 1/C1 + 1/C2 + ...1/Cn$$

Capacitors in parallel are calculated the same as resistors in series. Each capacitor increases the total surface area holding a charge. The two capacitances are simply added together.

$$CTOTAL = C1 + C2 + ...Cn$$

APPENDIX A
METRIC PREFIXES

Prefix	Symbol	Multiplier	Exponent
yotta	Y	1,000,000,000,000,000,000,000,000	10^{24}
zetta	Z	1,000,000,000,000,000,000,000	10^{21}
Exa	E	1,000,000,000,000,000,000	10^{18}
peta	P	1,000,000,000,000,000	10^{15}
Tera	T	1,000,000,000,000	10^{12}
giga	G	1,000,000,000	10^{9}
mega	M	1,000,000	10^{6}
Kilo	k	1,000	10^{3}
hecto	h	100	10^{2}
deca	da	10	10^{1}
1	unit	1	10^{0}
deci	d	0.1	10^{-1}
centi	c	0.01	10^{-2}
milli	m	0.001	10^{-3}
micro	μ	0.000001	10^{-6}
nano	n	0.000000001	10^{-9}
pico	p	0.000000000001	10^{-12}
femto	F	0.000000000000001	10^{-15}
Atto	A	0.000000000000000001	10^{-18}
zepto	Z	0.000000000000000000001	10^{-21}
yocto	Y	0.000000000000000000000001	10^{-24}

APPENDIX B
AMERICAN WIRE GAUGE

Please note that all values have been rounded, this table is meant as a guide - not an authority - for copper wire. Please check manufacturer data sheets when creating final designs for your applications.

* Max Current in Chassis Wiring

** Max Current in Power Transmission

Gauge #	Dia. (in.)	Dia. (mm)	XS Area (mm²)	Ω / 1000 ft.	Ω / km	Max I*	Max I**
0000	0.4600	11.6840	107.00	0.049	0.1607	380	302
000	0.4096	10.4038	84.90	0.062	0.2027	328	239
00	0.3648	9.2659	67.40	0.0779	0.2555	283	190
0	0.3249	8.2525	53.50	0.0983	0.3224	245	150
1	0.2893	7.3482	42.40	0.1239	0.4064	211	119
2	0.2576	6.5430	33.60	0.1563	0.5127	181	94
3	0.2294	5.8268	26.70	0.1970	0.6462	158	75
4	0.2043	5.1892	21.10	0.2485	0.8151	135	60
5	0.1819	4.6203	16.80	0.3133	1.0276	118	47
6	0.1620	4.1148	13.30	0.3951	1.2959	101	37
7	0.1443	3.6652	10.60	0.4982	1.6341	89	30
8	0.1285	3.2639	8.37	0.6282	2.0605	73	24
9	0.1144	2.9058	6.63	0.7921	2.5981	64	19
10	0.1019	2.5883	5.26	0.9989	3.2764	55	15
11	0.0907	2.3038	4.17	1.2600	4.1328	47	12
12	0.0808	2.0523	3.31	1.5880	5.2086	41	9.3
13	0.0720	1.8288	2.63	2.0030	6.5698	35	7.4
14	0.0641	1.6281	2.08	2.5250	8.2820	32	5.9
15	0.0571	1.4503	1.65	3.1840	10.4435	28	4.7
16	0.0508	1.2903	1.31	4.0160	13.1725	22	3.7
17	0.0453	1.1506	1.04	5.0640	16.6099	19	2.9
18	0.0403	1.0236	0.82	6.3850	20.9428	16	2.3
19	0.0359	0.9119	0.65	8.0500	26.4073	14	1.8
20	0.0320	0.8128	0.52	10.1500	33.2920	11	1.5
21	0.0285	0.7239	0.41	12.8000	41.9840	9	1.2
22	0.0253	0.6452	0.33	16.1400	52.9392	7	0.9

APPENDIX B
AMERICAN WIRE GAUGE

Please note that all values have been rounded, this table is meant as a guide - not an authority - for copper wire. Please check manufacturer data sheets when creating final designs for your applications.

* Max Current in Chassis Wiring

** Max Current in Power Transmission

Gauge #	Dia. (in.)	Dia. (mm)	XS Area (mm2)	Ω / 1000 ft.	Ω / km	Max I*	Max I**
23	0.0226	0.5740	0.2590	20.36	66.7808	4.70	0.7290
24	0.0201	0.5105	0.2050	25.67	84.1976	3.50	0.5770
25	0.0179	0.4547	0.1620	32.37	106.1736	2.70	0.4570
26	0.0159	0.4039	0.1280	40.81	133.8568	2.20	0.3610
27	0.0142	0.3607	0.1020	51.47	168.8216	1.70	0.2880
28	0.0126	0.3200	0.0800	64.90	212.8720	1.40	0.2260
29	0.0113	0.2870	0.0647	81.83	268.4024	1.20	0.1820
30	0.0100	0.2540	0.0507	103.20	338.4960	0.86	0.1420
31	0.0089	0.2261	0.0401	130.10	426.7280	0.70	0.1130
32	0.0080	0.2032	0.0324	164.10	538.2480	0.53	0.0910
M 2.0	0.0079	0.2000	0.0314	169.39	555.6100	0.51	0.0880
33	0.0071	0.1803	0.0255	206.90	678.6320	0.43	0.0720
M 1.8	0.0071	0.1800	0.0254	207.50	680.5500	0.43	0.0720
34	0.0063	0.1600	0.0201	260.90	855.7520	0.33	0.0560
M 1.6	0.0063	0.1600	0.0201	260.90	855.7520	0.33	0.0560
35	0.0056	0.1422	0.0159	329.00	1079.1200	0.27	0.0440
M 1.4	0.0055	0.1400	0.0154	339.00	1114.0000	0.26	0.0430
36	0.0050	0.1270	0.0127	414.80	1360.0000	0.21	0.0350
M 1.25	0.0049	0.1250	0.0123	428.20	1404.0000	0.20	0.0340
37	0.0045	0.1143	0.0103	523.10	1715.0000	0.17	0.0289
M 1.12	0.0044	0.1120	0.0099	533.80	1750.0000	0.16	0.0277
38	0.0040	0.1016	0.0081	659.60	2163.0000	0.13	0.0228
M 1	0.0039	0.1000	0.0079	670.20	2198.0000	0.13	0.0225
39	0.0035	0.0889	0.0062	831.80	2728.0000	0.11	0.0175
40	0.0031	0.0787	0.0049	1049.00	3440.0000	0.09	0.0137

APPENDIX C
RESISTOR COLOUR CODES

Colour	Value	Multiplier	Tolerance
Black	0	0	
Brown	1	10	± 1%
Red	2	100	± 2%
Orange	3	1,000	
Yellow	4	10,000	
Green	5	100,000	± 0.50%
Blue	6	1,000,000	± 0.25%
Violet	7	10,000,000	± 0.10%
Grey	8		± 0.05%
White	9		
Gold		0.1	± 5%
Silver		0.01	± 10%
None			± 20%

APPENDIX D
SCHEMATIC SYMBOLS

Appendix D – Schematic Symbols

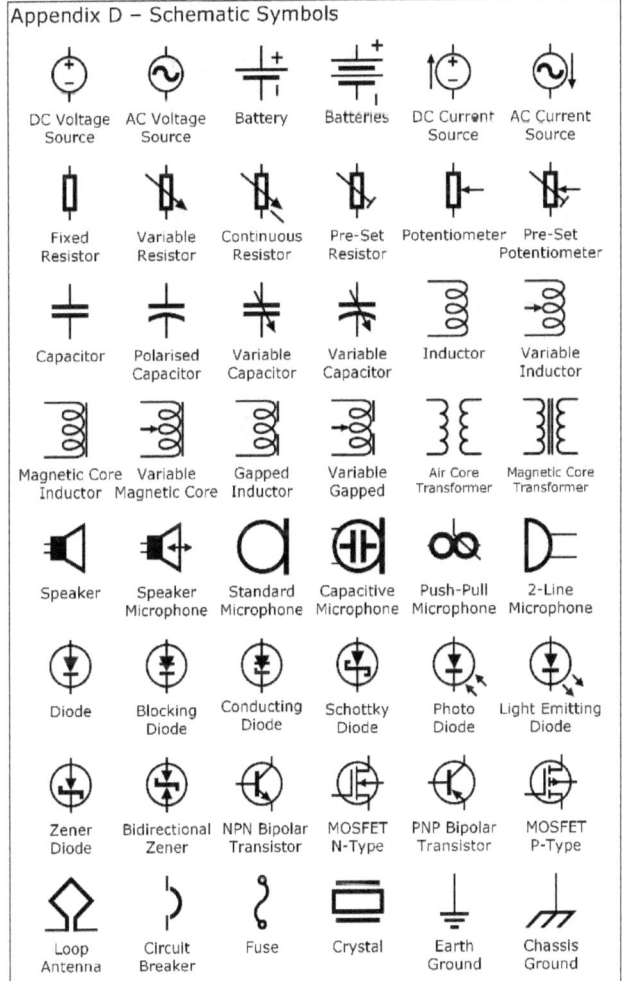

DC Voltage Source	AC Voltage Source	Battery	Batteries	DC Current Source	AC Current Source
Fixed Resistor	Variable Resistor	Continuous Resistor	Pre-Set Resistor	Potentiometer	Pre-Set Potentiometer
Capacitor	Polarised Capacitor	Variable Capacitor	Variable Capacitor	Inductor	Variable Inductor
Magnetic Core Inductor	Variable Magnetic Core	Gapped Inductor	Variable Gapped	Air Core Transformer	Magnetic Core Transformer
Speaker	Speaker Microphone	Standard Microphone	Capacitive Microphone	Push-Pull Microphone	2-Line Microphone
Diode	Blocking Diode	Conducting Diode	Schottky Diode	Photo Diode	Light Emitting Diode
Zener Diode	Bidirectional Zener	NPN Bipolar Transistor	MOSFET N-Type	PNP Bipolar Transistor	MOSFET P-Type
Loop Antenna	Circuit Breaker	Fuse	Crystal	Earth Ground	Chassis Ground

APPENDIX D
SCHEMATIC SYMBOLS

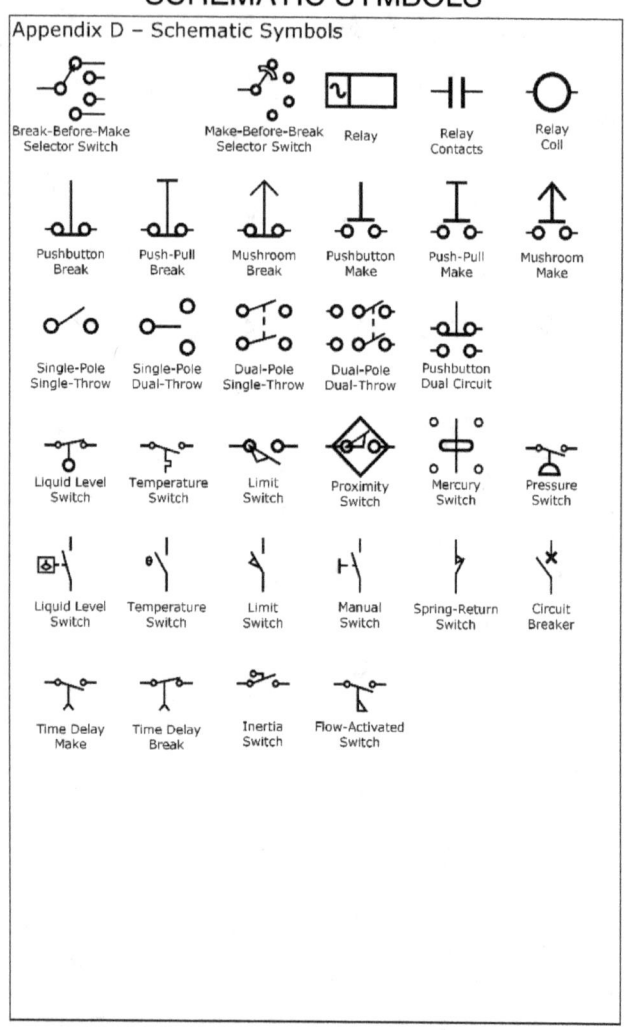

Appendix D – Schematic Symbols

Break-Before-Make Selector Switch

Make-Before-Break Selector Switch

Relay

Relay Contacts

Relay Coil

Pushbutton Break

Push-Pull Break

Mushroom Break

Pushbutton Make

Push-Pull Make

Mushroom Make

Single-Pole Single-Throw

Single-Pole Dual-Throw

Dual-Pole Single-Throw

Dual-Pole Dual-Throw

Pushbutton Dual Circuit

Liquid Level Switch

Temperature Switch

Limit Switch

Proximity Switch

Mercury Switch

Pressure Switch

Liquid Level Switch

Temperature Switch

Limit Switch

Manual Switch

Spring-Return Switch

Circuit Breaker

Time Delay Make

Time Delay Break

Inertia Switch

Flow-Activated Switch